ROCKS & LANDSCAPES OF THE NORTH YORK MOORS

Rocks and Landscapes of the North York Moors Pocket Edition
First published by High Tide Publishing 2018

ISBN 978-0-993-36462-4

Original illustrations Alan Marshall
Illustrations are based on sections produced by the British Geological Survey, used with permission kindly granted by the BGS

Designed by Woodenark, Scarborough
Printed by Adverset, Scarborough

A CIP record for this book is available from the British Library

Photograph of the Bridestones on p53 copyright Rich McGuinn, reproduced with kind permission

High Tide Publishing books are available to order at:
www.hightidepublishing.co.uk/shop

Retailers wishing to order stock should email: hightidepublishing@btinternet.com

ROCKS & LANDSCAPES OF THE NORTH YORK MOORS

Roger Osborne

Contents

Rocks and Landscapes of the North York Moors

Welcome to the North York Moors, one of the most spectacular and fascinating landscapes in Britain. This special guidebook shows how underlying geology has combined with forces of ice and water to produce such stunning effects. Knowing more about how the rocks and landscape fit together will make your experience of this beautiful landscape even more enjoyable. This book uses photographs and illustrations to guide general readers to a richer understanding of the moors and dales. Take our guide with you for a richer experience of a very special place.

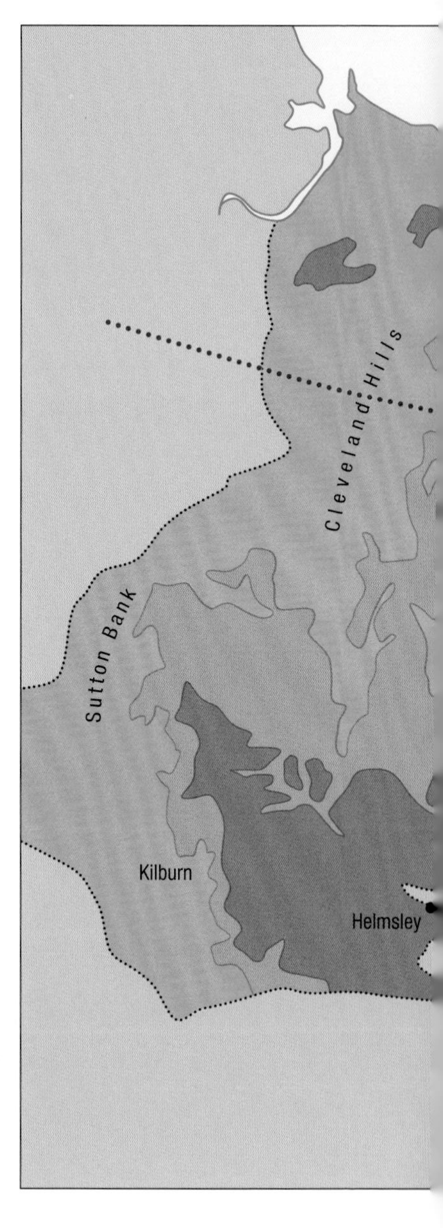

▶ **The rocks of the North York Moors become steadily younger as you travel south. The changing conditions has given us a huge variety of rocks. The variety of rocks is matched by the abundance and range of fossils. On each page we show a fossil that occurs in nearby rocks.**

The flat plain of the Tees has Triassic sediments buried under recent river deposits. The great plateau of the moors is made of Jurassic rocks. In the north, in the dales and along the coast are shales and ironstones formed in the seas of the Early Jurassic period. On top of them, making the hard crust of the moors are sandstones and gritstones of the Middle Jurassic, formed in a huge coastal delta caused by a drop in sea levels. In the Late Jurassic sea levels rose again; the rocks along the Tabular hills in the south are limestones and sandstones formed in a warm shallow sea. Further south again are beds of clay followed by the chalk hills of the Wolds, all formed in the Cretaceous period which came after the Jurassic.

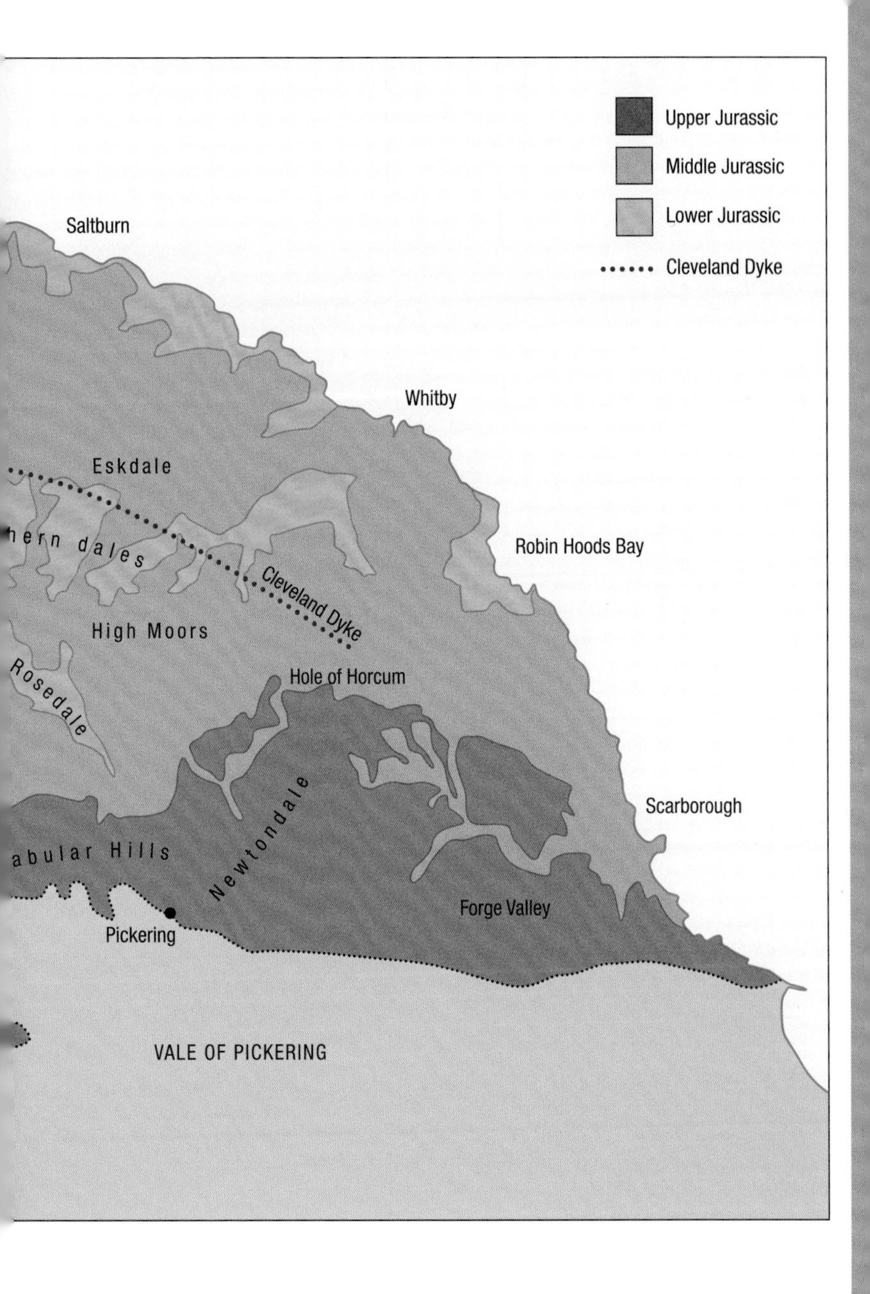
Upper Jurassic
Middle Jurassic
Lower Jurassic
Cleveland Dyke
Saltburn
Whitby
Eskdale
hern dales
Cleveland Dyke
Robin Hoods Bay
High Moors
Rosedale
Hole of Horcum
Newtondale
Scarborough
abular Hills
Forge Valley
Pickering
VALE OF PICKERING

Rocks beneath the moors

The North York Moors are built on a stack of Jurassic rocks that slope gradually towards the south. The key rock bed is the Middle Jurassic sand and gritstone (180 to 150 million years old), marked as green on the section, which lies beneath the central and northern moorland and the Cleveland Hills.

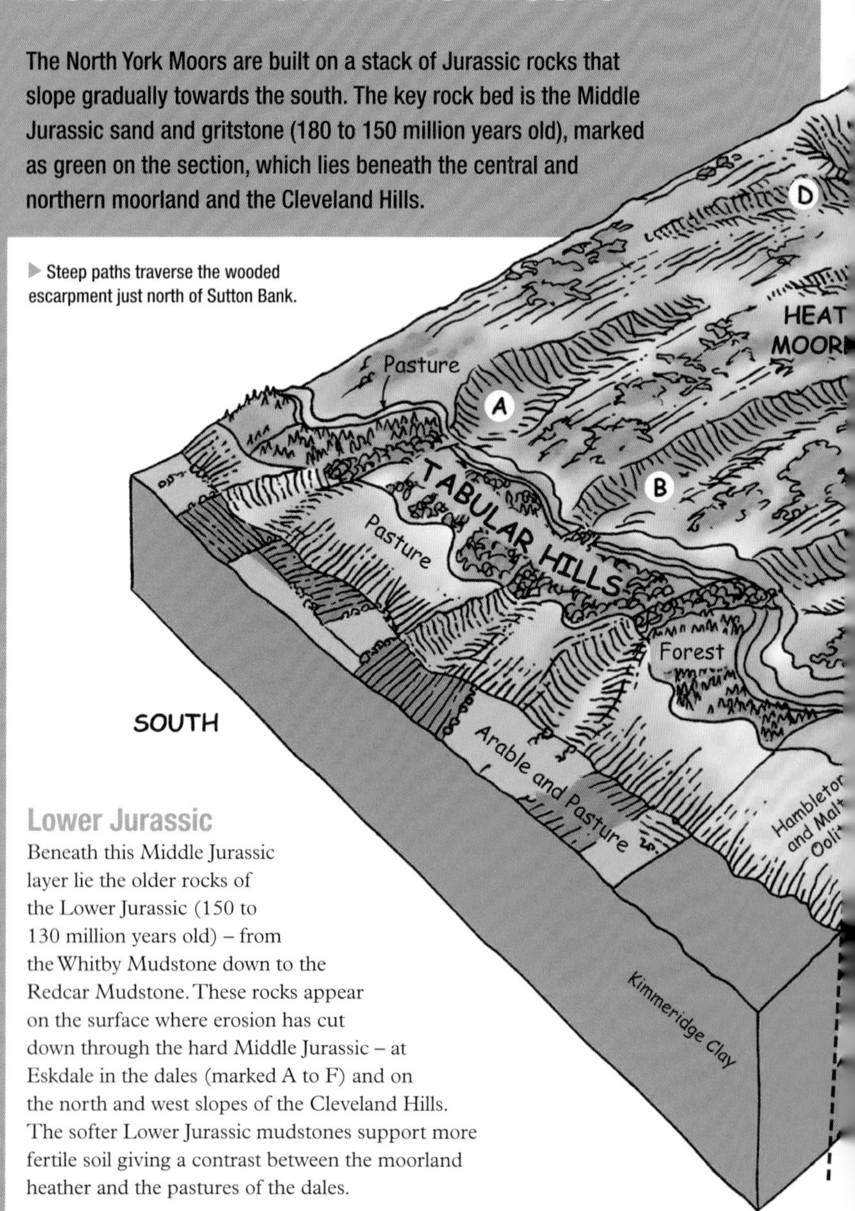

▶ Steep paths traverse the wooded escarpment just north of Sutton Bank.

Lower Jurassic

Beneath this Middle Jurassic layer lie the older rocks of the Lower Jurassic (150 to 130 million years old) – from the Whitby Mudstone down to the Redcar Mudstone. These rocks appear on the surface where erosion has cut down through the hard Middle Jurassic – at Eskdale in the dales (marked A to F) and on the north and west slopes of the Cleveland Hills. The softer Lower Jurassic mudstones support more fertile soil giving a contrast between the moorland heather and the pastures of the dales.

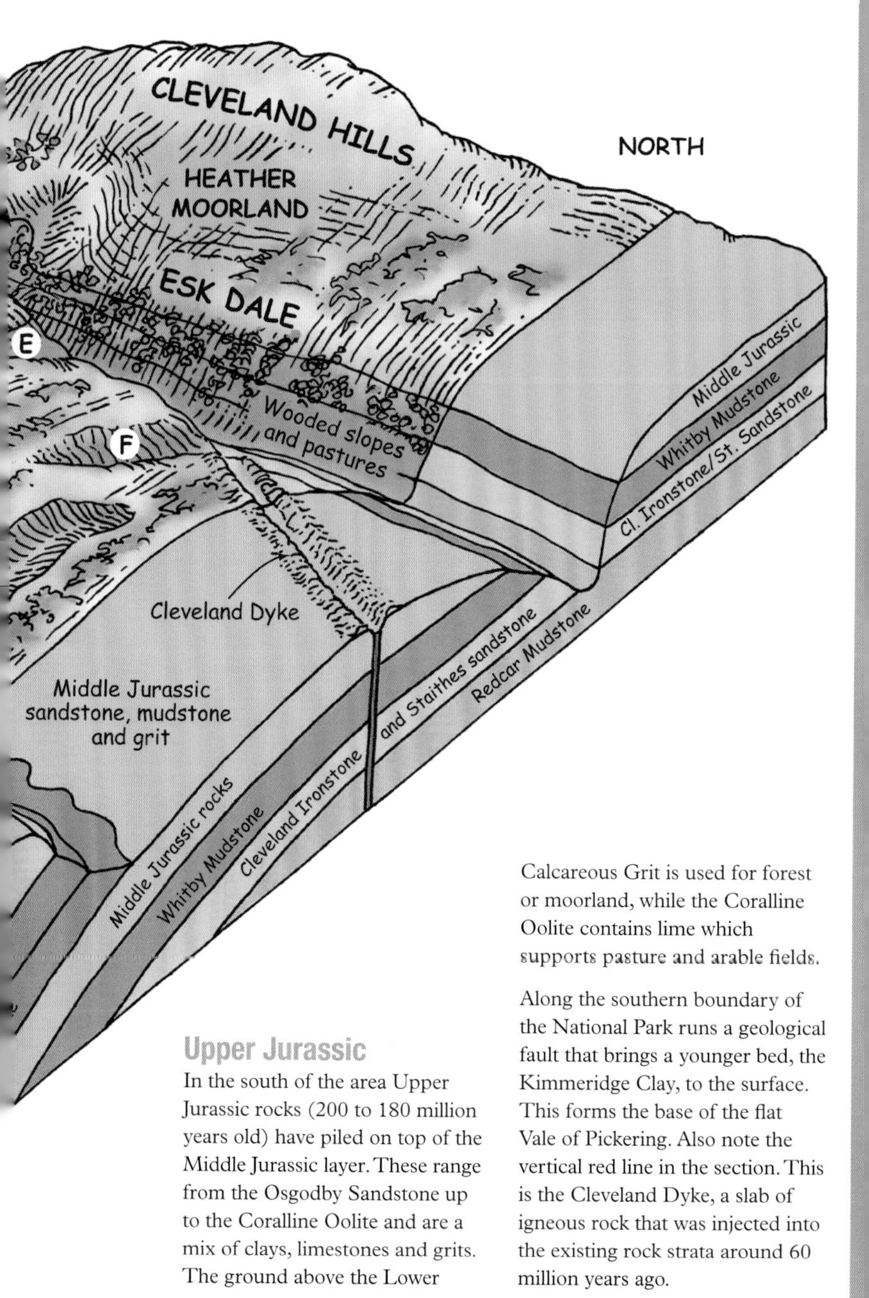

Upper Jurassic

In the south of the area Upper Jurassic rocks (200 to 180 million years old) have piled on top of the Middle Jurassic layer. These range from the Osgodby Sandstone up to the Coralline Oolite and are a mix of clays, limestones and grits. The ground above the Lower Calcareous Grit is used for forest or moorland, while the Coralline Oolite contains lime which supports pasture and arable fields.

Along the southern boundary of the National Park runs a geological fault that brings a younger bed, the Kimmeridge Clay, to the surface. This forms the base of the flat Vale of Pickering. Also note the vertical red line in the section. This is the Cleveland Dyke, a slab of igneous rock that was injected into the existing rock strata around 60 million years ago.

The White Horse at Kilburn marks the steep escarpment at the western edge of the North York Moors. The moors sit on a plateau standing above the surrounding countryside. Escarpments often mark geological boundaries and are generally covered in forest.

Sutton Bank

Sutton Bank is a wonderful place to start our exploration of the rocks and landscapes of the North York Moors. The western escarpment marks the edge of the National Park and is one of the most spectacular inland cliffs in England. The cliff is there because the land to the east – the upland area of the moors – was lifted up during the last fifty million years. The Vale of Mowbray was then invaded by a series of ice sheets. The edges of ice sheets are packed with rock fragments and have immense power to grind through the surrounding land. Here the ice sheet sheered off a near-vertical cliff around 200 metres high as it travelled south.

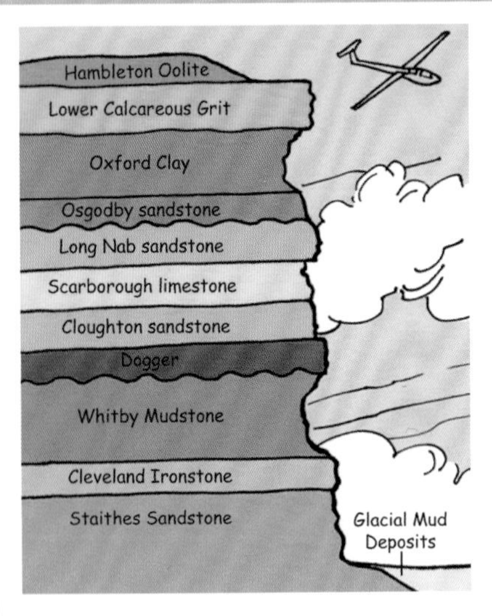

The cliff at Sutton Bank shows a slice through almost the whole Jurassic sequence of rocks. The Hambleton Oolite outcrops on the plateau above, but the cliff top itself is a massive slab of hard Lower Calcareous Grit. The cliff is so steep because this hard cap sits on top of softer rocks that get eaten away from underneath the grit. This means that falls of boulders from the higher harder beds are common. The Oxford Clay and Whitby Mudstone are particularly soft.

The abandoned quarries at Boltby Scar expose thick layers of Lower Calcareous Grit. This blocky stone, which dates from the Late Jurassic, is used for building material all across the moors.

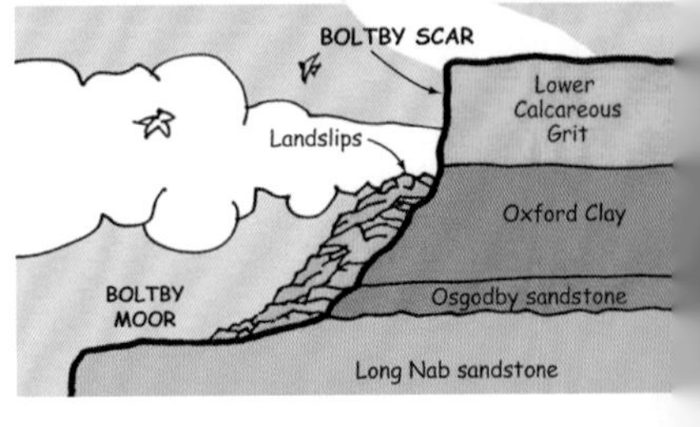

Boltby Scar

Just to the north of Sutton Bank is a magnificent exposure of Upper Jurassic rock at Boltby Scar. Here the hard layer of Lower Calcareous Grit has been undermined by erosion, leading to a series of huge landslips or 'tumbledowns' and leaving this cliff face.

Boltby Moor, which is a shelf of rock sticking out at a lower level, is made of the next hard layer down – the Long Nab Grit. It is likely that a portion of the ice sheet pushed over this lower layer.

A massive slab of hard gritstone lies on the top of Sutton Bank as seen here at Roulston Scar. As the softer rock beneath gets eaten away, massive grit boulders tumble down the hillside.

Gormire Lake

One of the few natural lakes in Yorkshire, Gormire is a remnant of the last ice age. As the ice sheet melted boulder clay and water was strewn across the landscape. A bank of clay has blocked natural drainage routes, creating a lake.

▼ Steep paths traverse the wooded escarpment just north of Sutton Bank.

Bourgetia

A large marine gastropod, related to snails, lived on the floor of shallow Upper Jurassic seas. Gastropods are common throughout the fossil record. In this area they occur in Lower Jurassic and Upper Jurassic strata. Often they are found in section, and occasionally as whole specimens like this.

Bilsdale and the Hambleton Hills

The western portion of the North York Moors is cut through by the beautiful valley of Bilsdale, running south from the edge of the Cleveland Hills towards Rievaulx. To the west lies the open moorland of Snilesworth Moor and Hambleton Moor. But a little further south lie the Hambleton Hills, which are entirely different in their geology and landscape. This range of hills is formed from Upper Jurassic rocks laid on top of the Middle Jurassic grits and sandstones of the moors.

The river Seph runs from north to south through Bilsdale. The highest land in the National Park (454 metres above sea level) is on Urra Moor just to the east of the dale's northern end. Bilsdale is formed by the river cutting through the hard cap of Long Nab gritstone down into the soft Whitby Mudstone beneath. The valley cuts down through the layers of ironstone and sandstone to the Redcar Mudstone in the valley floor.

▼ This idealised section across the western part of the North York Moors shows how Bilsdale and the Hambleton Hills are related. The Upper Jurassic rocks of the Hambleton Hills are stacked on top of the Middle Jurassic gritstone of the main plateau of the North York Moors. The great conical hills at Coomb, Hawnby and Easterside are outliers from the Hambleton Hills, topped by a layer of Hambleton Oolite.

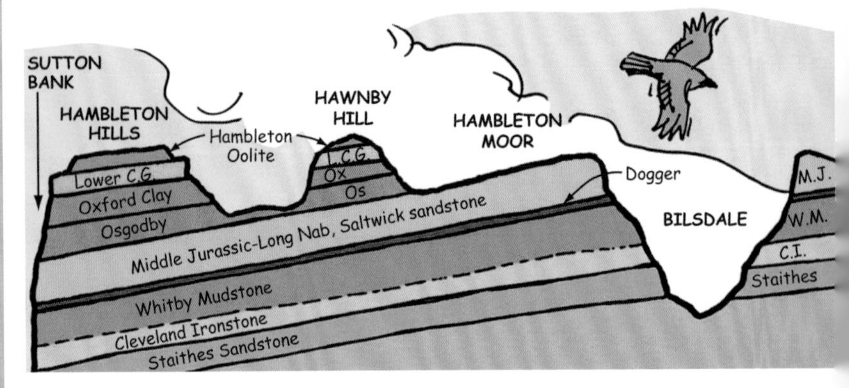

Reading the landscape

The road from Helmsley north towards Stokesley rises gradually over the southerly dip of the Upper Jurassic rocks of the Tabular Hills. It then makes a sudden spectacular dive down Newton Bank into Bilsdale. This escarpment marks the change to Middle Jurassic rocks. The road rises once again going north through Chop Gate, now climbing the southerly dip of the Middle Jurassic. The climb comes to a sudden halt with a steep descent over Clay Bank, which marks the end of the Middle Jurassic, giving spectacular views over the Tees floodplain.

▲ The cap of Easterside hill is a thin layer of Hambleton Oolite on top of hard Lower Calcareous Grit. This tough rock makes the top of the hill infertile, while the softer Oxford Clay on the lower slopes supports pasture fields.

Cardioceras

This beautiful ammonite fossil is found in Upper Jurassic limestones. Ammonites floated freely in the Jurassic seas, using the different chambers in their shells for buoyancy. The shells fell to the sea floor when they died.

Kilburn and the southwest

The southwest margin of the North York Moors has both beautiful landscapes and fascinating geology. Just like at Sutton Bank, the uplands come to a sudden stop at a steep cliff overlooking the area beneath the famous White Horse. But the history of this landscape is a little different, as the cliffs have been created by a series of faults. These faults have isolated the Howardian Hills to the south by creating the lowland channel known to geologists as the Coxwold-Gilling Gap.

While it is a magnificent sight, the White Horse of Kilburn is a double pretence. Inspired by the prehistoric white horse at Uffington, Victorian enthusiasts cut horses on chalk hillsides across England. The Kilburn horse, Britain's largest, was cut in 1857. The underlying rock is Upper Calcareous Grit, so chalk has been regularly brought to the site to keep the White Horse white.

The Hambleton Hills, Tabular Hills and Howardian Hills (below) are mainly made of Upper Jurassic rocks, while the Wolds to the southeast are made from even younger chalk. The Coxwold-Gilling Gap links the Vale of York to the Vale of Pickering, a flat lowland vale surrounded by areas of upland. The Vale of York glacier that pushed south between the North York Moors and the Pennines sent a spur into the gap, blocking off drainage from the Vale of Pickering, which was then a shallow lake.

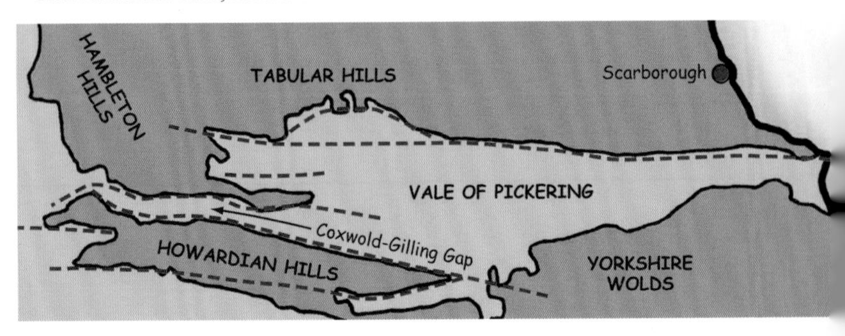

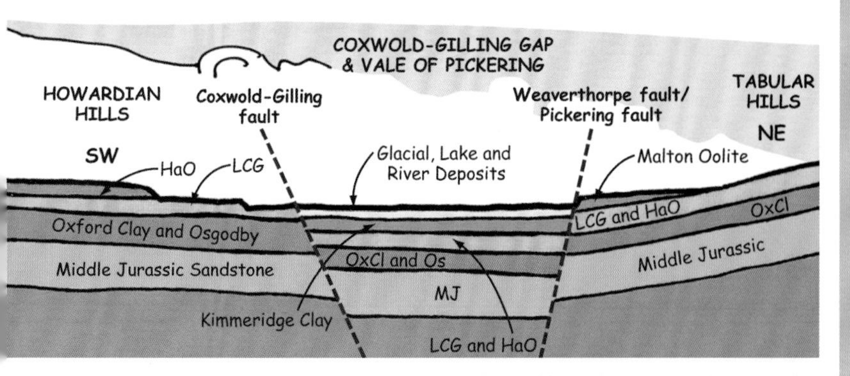

Coxwold-Gilling Gap

A complex system of geological faults created a channel of low lying land between the Hambleton Hills and the Howardian Hills to the south. The faulting happened mostly in the Tertiary period, long after these Jurassic rocks had been formed and uplifted. The faults generally run from west to east and the land between them has dropped to create a flat valley bottom, now covered in recent mud deposits.

Pleuromya

This bivalve occurs in Upper Jurassic marine strata. It is around 50 mm across. Bivalves are like present-day mussels, attached to the sea floor or rock surfaces, opening and closing their shells to regulate the flow of sea water, while feeding off microscopic nutrients.

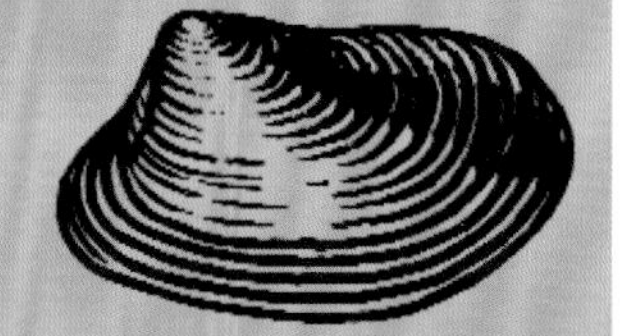

▲ Lime-rich rocks give the Howardian Hills, to the southwest of the moors, an abundance of rolling pastures and broad-leaved copses.

The majestic peak of Roseberry Topping stands high above the Tees floodplain. The hill is an outlier of the Cleveland Hills, capped by a hard piece of the same Middle Jurassic grit that forms the great escarpment to the south. Once that cap is eroded away the hill will be worn down to the level of the surrounding countryside.

Cleveland Hills

The Cleveland Hills form a magnificent arc that marks the boundary between the uplands of the North York Moors and the low-lying Tees Valley. The hard rock (Middle Jurassic sandstone, mudstone and limestone) of the moorland slopes gradually upwards towards the north and west. But this rock is being slowly worn away at its northern edge, forming a steep escarpment. This gives breathtaking views over the Tees Valley. Seen from below, the hills are an imposing gateway to the moors.

A view of a section of the Cleveland Hills, looking southwest from Easby Moor. On the flat floodplain below are the villages of Great Broughton and Kirkby, with Carlton Bank in the distance.

▲ Cutting in Guisborough Forest revealing alum shale.

▲ A pile of alum slag in woods above Guisborough.

Alum

Guisborough Forest runs along the northern escarpment of the Cleveland Hills. Its upper levels are marked by the remains of huge alum quarries. The Alum Shale is part of the Whitby Mudstone Formation. The alum industry quarried vast amounts of shale from the Cleveland Hills and the Yorkshire coast from the 1600s until around 1870. The shale was roasted in huge clamps before the alum – used in the textile trade that was the basis of the industrial revolution – was crystallised out of the resulting liquor.

The first alum quarries were at Slapewath and Carlton Bank. The focus of the industry later moved to the coast, where access and transport were easier. Mounds like these in Guisborough Forest are made of piles of left-over alum shale.

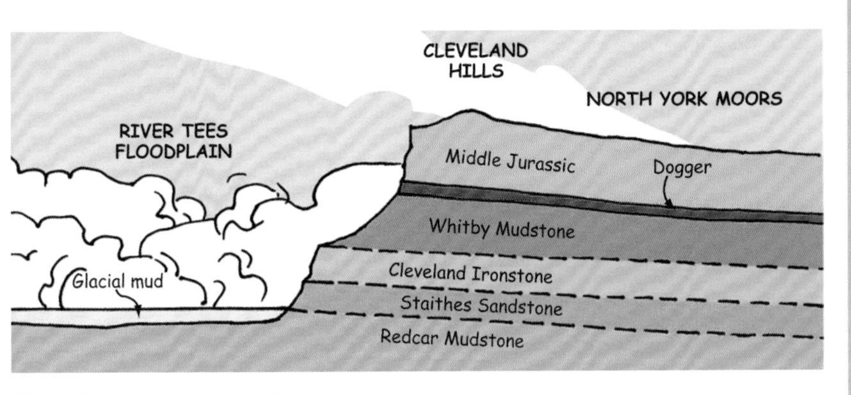

Cleveland escarpment

Middle Jurassic rocks that lie beneath the high moorland of the North York Moors form the flat summits of the Cleveland Hills. The softer mudstones underneath are continually eaten away causing the hard rock to break off giving a steep northward slope. The Whitby Mudstone and Cleveland Ironstone beds have been the source of mineral wealth in this area for centuries.

This view of the Cleveland Hills escarpment above Battersby shows a rocky crag at the top. This is an exposure of Long Nab sandstone. The lower slopes are peppered with boulders of sandstone, while further down the Whitby Mudstone supports more fertile soil for pasture.

Ichthyosaurs

Ichthyosaurs have been found in different beds within the Whitby Mudstone; this is a section of backbone from an ichthyosaur (below).

Roseberry Topping

This remarkable hill is the best-known landmark in the lower Tees Valley. A hard slab of sandstone sits on top of Roseberry Topping, while the soft Whitby Mudstone is eaten away from beneath. The hill also owes its shape to mining – ironstone on the lower slopes was dug out in the past, leading to the collapse of one side of the hill. The climb to the top gives spectacular views over Teesside and along the Cleveland Hills.

Roseberry Topping is an outlier of the Cleveland Hills. The hard layer that sits on top of the hills is being slowly eroded from the northern and western edge of the hills. The small piece on Roseberry Topping is a remnant of that erosion that will eventually disappear.

Because the rock strata tilt upwards towards the north, the northern edge is being continually worn away.

Geology is a continuous process - the material worn off these hills is eventually washed into rivers and seas, where it is laid down to form future rocks.

▼ **A hard cap of Middle Jurassic sandstone overlays a thick layer of soft Lower Jurassic Whitby Mudstone. Erosion has isolated the hill from the rest of the Cleveland Hills, which have identical rock formations.**

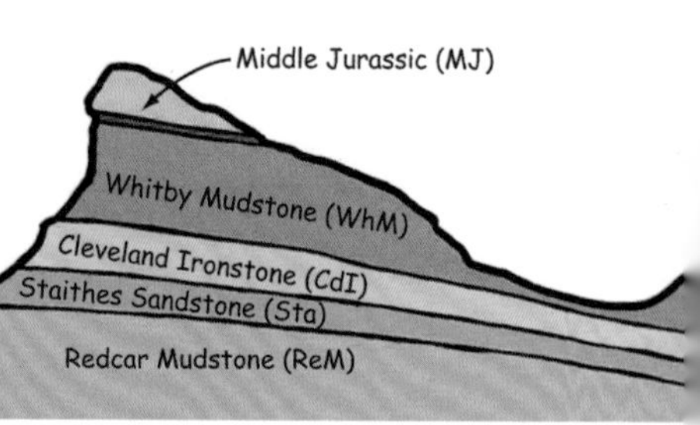

Hard and soft rocks

The west face of Roseberry Topping is a massive cliff of yellow moorland sandstone from the Saltwick Formation. This same type of rock forms the plateau of the North York Moors stretching south as far as the Tabular Hills. The soft rocks beneath are worn away by rain and wind, allowing huge boulders of hard sandstone to break off and tumble down. Look out for these boulders on the slopes below Roseberry Topping.

▲ This face of Roseberry Topping shows Middle Jurassic sandstone.

Eston Hills

The Eston Hills comprise another outlier of the Cleveland Hills. They are separated by a wide flat valley in which the town of Guisborough sits. The valley is a small rift valley or graben – the central section has slipped down the faults on either side; the valley floor is covered in glacial deposits.

The Cleveland Way follows the steep Cleveland Hills escarpment where rocky crags of sandstone are undercut by erosion of the soft shales beneath. The sandstone is eroded into spectacular shapes at the Wainstones, probably by ice and wind.

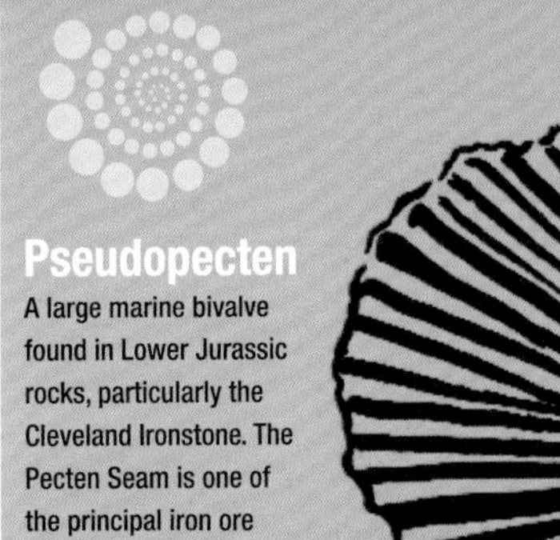

Pseudopecten

A large marine bivalve found in Lower Jurassic rocks, particularly the Cleveland Ironstone. The Pecten Seam is one of the principal iron ore seams in Cleveland.

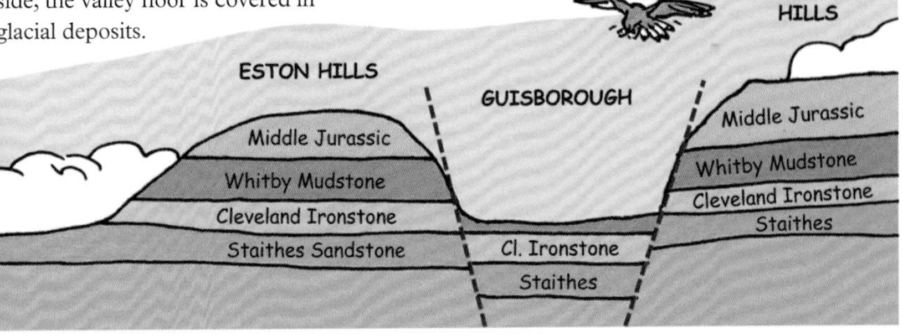

By the time it reaches Larpool just upstream from Whitby, the Esk has cut a steep gorge, shown by this view from Larpool railway viaduct. The steep sides of the gorge here and at Whitby show that the Esk's present course is relatively new, probably created after the last ice age.

Upper Eskdale

The River Esk begins its 50-kilometre journey to the sea in Westerdale, before flowing into the broad open expanse of upper Eskdale. The beautiful landscape of upper Eskdale has one peculiarity which comes from its recent geological history. We might expect an upland stream to create a narrow valley, and then gradually open out into a broad floodplain. But the Esk does the opposite – upper Eskdale is broad and quite flat, while lower Eskdale is steep and narrow. The cause of this upside-down dale can be found in the last ice age.

▼ Ice sheets pushed in from the North Sea and down the Vale of York but were not thick enough to cover the high ground of the North York Moors. One piece of the North Sea ice pushed up into Eskdale, probably as far as Lealholm. This wall of ice blocked all the drainage from the moors to the sea and created a glacial lake. A layer of mud was deposited on the floor of the lake giving upper Eskdale its level appearance. A bank of glacial mud known as a moraine also marks the furthest extent of the ice sheet near Lealholm.

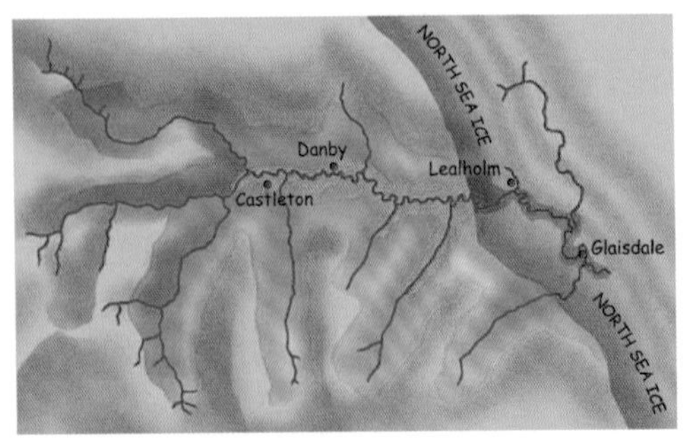

Lake Eskdale

Like the other dales that cut through the central part of the moors Eskdale is capped on either side by Middle Jurassic moorland sandstone (see section opposite). The Dogger marks the boundary between this and the Lower Jurassic rocks beneath.
As well as supporting pasture fields, the Whitby Mudstone Formation contains alum and cement shales as well as jet rock. With ironstone, whinstone from the Cleveland Dyke, and local clay for brick these natural resources made Eskdale a thriving industrial centre in the nineteenth century.

▲ From Castleton station to beyond Houlsyke the River Esk meanders gently through a flat open plain. The meanders, like this one near Danby, are cut through the old lake deposits from the last ice age, by a river flowing on a fairly level course.

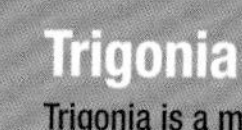

Trigonia

Trigonia is a marine bivalve found in rocks of the later Lower Jurassic period. Its characteristic shape is like a three-sided pyramid.

▶ The Esk broadens as it approaches Egton Bridge though the steep road down from Egton village to the river shows the depth of the gorge cut by the Esk.

River Eskdale

Like the other dales that cut through the central part of the moors Eskdale is capped on either side by Middle Jurassic moorland sandstone (see section below). The Dogger marks the boundary between this and the Lower Jurassic rocks beneath.

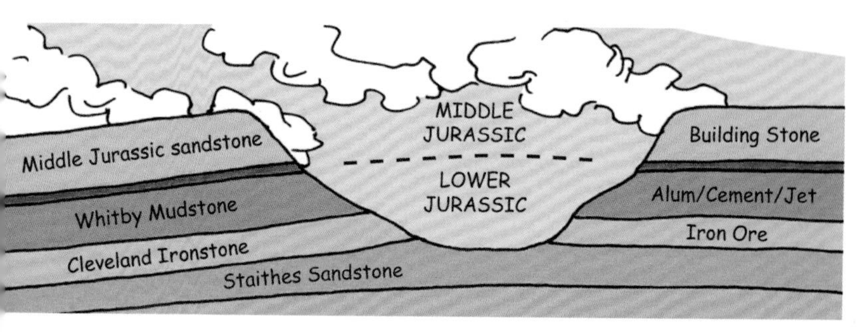

Lower Eskdale

The lower part of Eskdale is a beautiful and surprising landscape. From Glaisdale to Whitby the Esk has cut a series of steep sided gorges, behaving like a youthful mountain stream. But in its upper reaches around Danby and Lealhom the river winds gently through an open flat-bottomed valley. The reason for this upside down valley lies in its geological history. Lower Eskdale is actually a new river valley cut since the last ice age; while we expect a river to have a wide floodplain as it nears the coast, the deep gorge of the Esk runs all the way to Whitby.

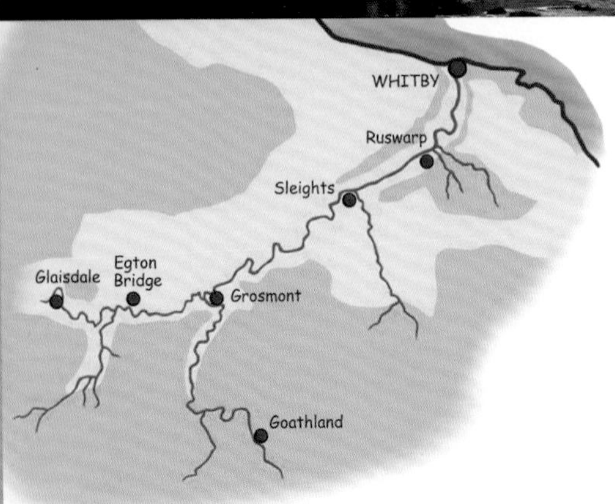

Esk gorge

The Esk enters a steep gorge at Limber Woods after it passes under the Beggar's Bridge at Glaisdale. If you ride the Esk valley railway from Whitby you pass through a series of gorges before emerging into open country west of Lealholm.
The hard Middle Jurassic rocks that form the tops of the moors press in close to the sides of Eskdale in its lower reaches.

◀ Access to the villages along the lower Esk is via steep winding roads. That's because the river is in a gorge; building roads along the river was always difficult if not impossible in most places.

By the time it reaches Sleights the Esk has cut a deep gorge. The road bridge is high above the river, and the north face of Eskdale is a steep wooded slope.

Dactylioceras

A beautiful ribbed ammonite commonly found in the Lower Jurassic rocks of the moors and coast. Dactylioceras specimens measure around 50mm across

▼ The lower Esk valley was an industrial hub through most of the nineteenth century. Ironstone was processed at Glaisdale and Grosmont before being taken away by train.

▲ The Esk Valley railway line, seen here near Ruswarp, follows the river through a beautiful changing landscape.

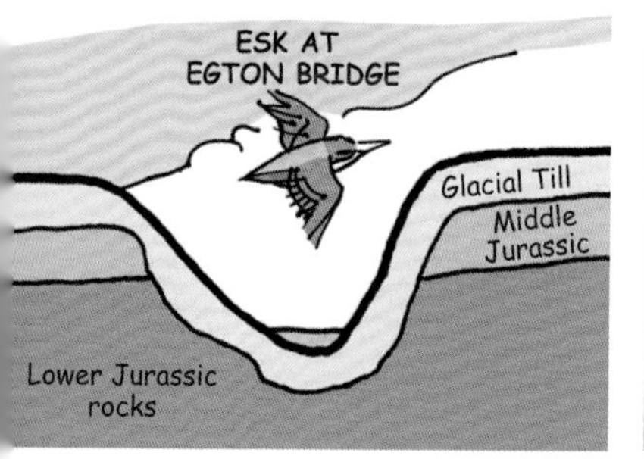

▲ At Egton Bridge the floor of Eskdale is covered in a yer of glacial till, showing that there was a valley here efore the ice age. The Esk has made this deeper and teeper.

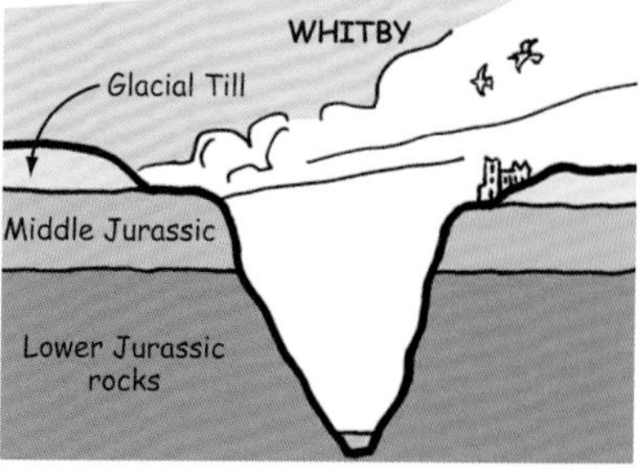

▲ Near Whitby the river seems to be in a completely new valley; it has stripped away the glacial till from the ground surface, and cut through the hard Middle Jurassic rock to scour a deep gorge.

Northern dales

A series of spectacular and beautiful dales runs from the high moorland down into Eskdale. Kildale and Commondale run southwards. Baysdale, Westerdale (which carries the Esk from its source), Danby Dale, Fryup Dale and Glaisdale run northwards, followed to the east by the dale of the Murk Esk and Iburndale. Each of these dales has been formed by a small stream cutting through the hard layer of Middle Jurassic rocks lying on top of the high plateau of the moors. Softer, more fertile Lower Jurassic rocks are exposed along the sides and floors of the dales, leading to a picturesque contrast between the moor tops and the rich green pastures of the dales.

Moorland and pasture

This view of the upper end of Fryupdale (above) shows a plateau separating the sweeping valleys of the two Fryupdales. Notice how this high ground is capped by heather and bracken, with pasture fields running down the side. This indicates a cap of hard moorland sandstone lying on top of softer more fertile shales (see section below). You will see this same pattern repeated in all the dales of the North York Moors.

The picture below shows a landscape known as The Hills; these are rounded hummocks formed by a combination of ice and landslips

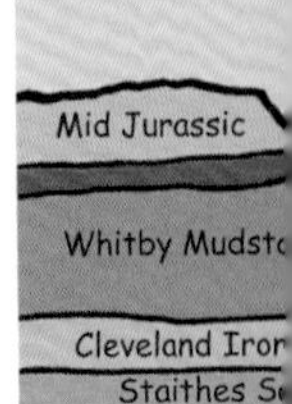

Ice and stone

The double dale of Fryup is separated into two by an outlier of hard Long Nab gritstone lying on top of the softer Whitby Mudstone. This forms a high piece of ground called the Heads. The ice sheet that pushed up the Esk valley came into Great Fryup and left behind a layer of glacial till. It did not reach up as far as Little Fryupdale, which has bedrock beneath its covering of soil.

The dales are divided into pasture fields by stone walls and hedges, with the pasture used for cattle, sheep and winter fodder.

Walls are used as field boundaries, marking the border between moorland and pasture, following the geological divide between the sandstone of the moorland and the mudstone.

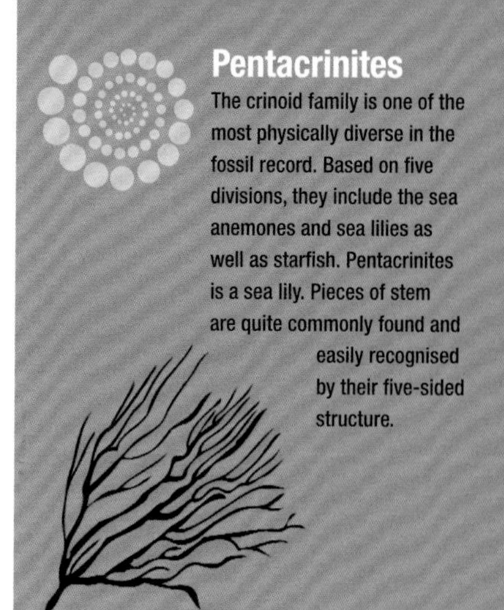

Pentacrinites

The crinoid family is one of the most physically diverse in the fossil record. Based on five divisions, they include the sea anemones and sea lilies as well as starfish. Pentacrinites is a sea lily. Pieces of stem are quite commonly found and easily recognised by their five-sided structure.

This barn at the head of Glaisdale shows the steepness of the dale side. The barn, like the field walls and farmhouses, is made of square blocks of the Middle Jurassic sandstone that covers much of the moors.

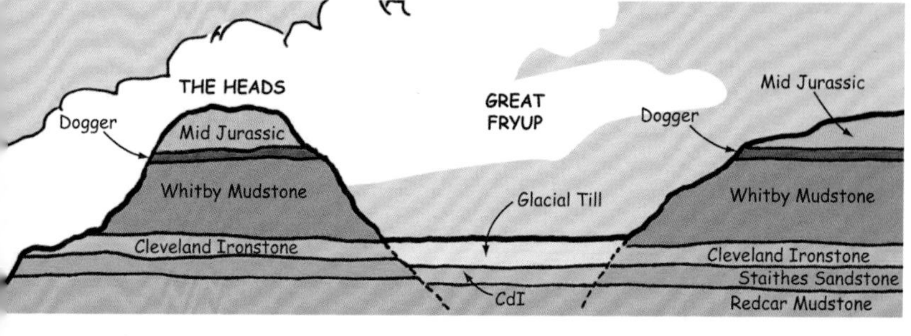

Highland cattle graze on the high moorland near the Hole of Horcum. Here the purple heather is mixed with rough grasses. Poor drainage gives good habitat for wading birds like curlew and golden plover with reed beds marking the locations of peat bogs.

High moors

The central feature of the North York Moors is the unbroken expanse of heather moorland that lies at its heart. The moorland has been created by a combination of natural environment and human activity dating back as far as the Bronze Age. In that time the landscape has changed from mixed woodland to a combination of peat bogs and heather moor; and all the time the main influence has been the underlying rocks. The rock beneath the high moorland is a thick layer of gritty sandstone formed in the Middle Jurassic period. On these pages we are going to look in a little more detail at those moorland rocks.

▲ The high moorland is a level plateau with poor drainage.

▲ The moors are peppered standing stones of different types and ages.

Human landscape

The heather moorlands are a human creation. In the warming that followed the end of the last ice age around 11,500 years ago the moors became covered in a mixed forest, firstly 'pioneer' species such as birch, willow and hazel, then later by oak, lime and alder. Humans came to the area during the Mesolithic era, from 10,000 to 6,000 years ago, and began burning and felling trees and undergrowth as part of a hunting strategy.

By 3,500 years ago the tree cover had mostly gone; soil nutrients leached out leaving a heathland vegetation.

▼ The few streams that cross the boggy moors are filled with boulders of hard Middle Jurassic rock.

▲ Deep layers of peat are visible where tracks have cut through the vegetation.

At Eller Beck great slabs of sandstone have been broken away and transported by the action of glacial meltwater and more recent floods.

Delta sandstones

In the last couple of centuries humans have intervened again and the heather moorland has been preserved as an ideal habitat for game birds. Each section is regularly burned off under controlled conditions.

All the rocks on the high moors are of Middle Jurassic age, formed when this region was a river delta on the coast of a subtropical sea. The lowest set of rocks here is the Cloughton Formation, a fine sandstone formed in the freshwater channels of the delta. This formation contains lots of plant fossils. Above that comes the Scarborough limestone, a set of marine rocks that shows that sea levels rose and flooded the area. Above the Scarborough limestone comes the Moor Grit, a hard gritstone, and then above that the Long Nab Member, of mudstone and sandstone.

The geology of the Middle Jurassic is a bit more complicated because of the environment. Deltas and coasts are always active, changing landscapes, with lots of different environments within them.

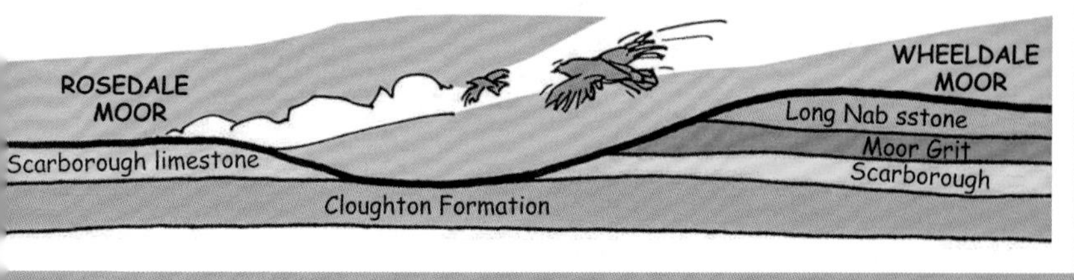

Dinosaur footprint

In the Middle Jurassic period dinosaurs roamed along the coast, with its rich vegetation. Few dinosaur bones have been found here, suggesting that conditions for fossilisation were not good. But dinosaur footprints are relatively plentiful, giving information about height, weight, claws and agility.

Moorland landforms

The high moorland owes its present appearance to the Middle Jurassic rocks that lie beneath and to human intervention – felling trees and burning off heather to preserve the vegetation. But the moorland is not a uniform landscape, and the variations in topography and vegetation give us more clues to its fascinating history.

▲ This view of the moorland road near Stape shows contrasting use of the infertile moorland. The area on the right is heather moorland; on the left recently planted conifer forest.

▼ This is Bella Dale Slack near Scaling Dam, a typical moorland slack. The drainage channel cut by meltwater is now filled by a bog.

Slacks and swangs

Ice sheets are complex environments dynamic and powerful with huge forces working on the landscape. Although frozen water makes up the bulk of the ice sheet, there are channels of running water beneath and around the ice, and when the ice melts vast amounts of water are released.

The moorland to the north and sout of Eskdale has a number of channels carved by glacial meltwater, known as slacks and swangs. They are quite distinct because they look like valleys carved by great rivers, but the rivers are no longer there, or are reduced t small becks.

A small herd of Highland cattle take advantage of the peat landscape of the high moors, cooling off in Seive Pond on the track from the Hole of Horcum towards Levisham Moor. The rough grazing perfectly suits these magnificent animals.

Astarte

Astarte is a fairly common but distinctive bivalve. The presence of marine fossils in a limestone formation shows the rise in sea levels in one interval of the generally land-dominated Middle Jurassic environment.

Moorland drainage

The even plateau of the moors has seen a build up of peat. This is created by vegetation decaying in the acidic conditions which are due in part to the underlying rock strata. The occasional slacks and swangs show up as shallow gulleys. They do not in general carry streams but are remnants of meltwater.

▼ Steep gulleys run down from the moorland plateau. The reed beds show the presence of standing water and peat bogs.

◀ This view shows the evenness of the plateau of the moors, allowing views over tens of miles of country.

Cleveland Dyke

A great landmark of the region, the Cleveland Dyke runs arrow straight across the northern part of the North York Moors. This is the only igneous rock formation in northeast Yorkshire; it forms a straight ridge, known locally as the Whinstone Ridge, running roughly from Blea Hill on the Fylingdales Moor in the east, across Eskdale and to the north. This is where the dyke reaches the surface; beneath ground it reaches out into the North Sea and up to the western isles of Scotland, forming part of a swarm of dykes. The Cleveland Dyke near Goathland is now an amazing sight as the hard rock has been quarried out, leaving a trough up to 20 metres deep.

The Cleveland Dyke near Goathland was originally a ridge caused by a hot slice of magma pushing up to the surface of the earth. While the sides of the ridge are still visible, the dyke is now a gigantic quarried trough, stretching across the high moors and Eskdale. Along the sides of the trough that marks the Cleveland Dyke there are signs of the local Jurassic rock being baked by the sudden injection of hot magma.

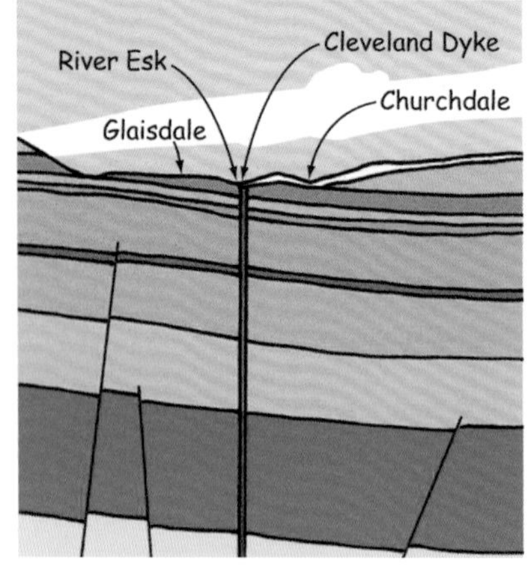

◀ This section (left) shows layers of rock deep beneath the surface. The Cleveland Dyke slices through these old rocks before reaching the surface on the North York Moors.

▼ The straight line of the Cleveland Dyke, marked in red on the map (below), crosses the boundaries between the other rocks, showing that it is a younger formation. The dyke was formed when a massive igneous event beneath western Scotland pushed hot magma outwards around 5 million years ago.

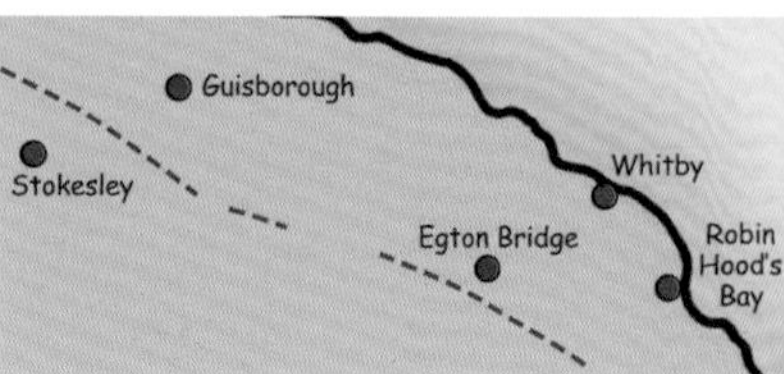

Dyke swarm

The Cleveland Dyke is the most southerly of a swarm of dykes created by underground igneous activity around 58 million years ago. A significant upswelling of magma was centred roughly beneath the Isle of Mull, around 350 kilometres away, though it may be that the Cleveland Dyke was pushed out of a reservoir of hot magma beneath the southern uplands of Scotland.

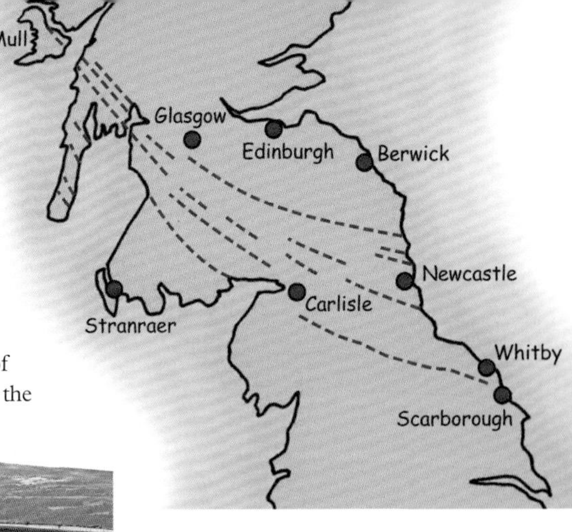

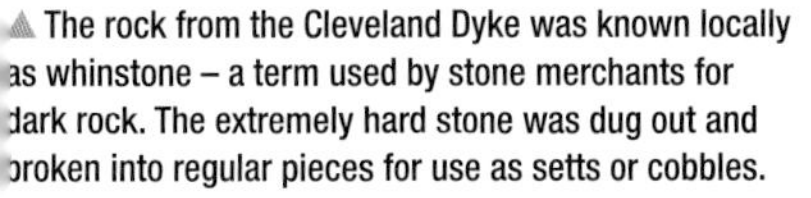

The rock from the Cleveland Dyke was known locally as whinstone – a term used by stone merchants for dark rock. The extremely hard stone was dug out and broken into regular pieces for use as setts or cobbles.

The vast quantities of stone taken out of the dyke leave traces of its origins. The sandstone margins of the dyke were super-heated by the igneous intrusion, leaving a layer of baked rock on the edges of the trough.

Cleveland Dyke dolerite

The rock of the Cleveland Dyke is a dolerite, a dark and fine-grained igneous rock type. The sedimentary rocks that dominate this region have been laid down in seas or deltas. The sandstones and mudstones are recycled rocks – sand and silt being worn away from existing rocks by seas and rivers. But the Cleveland Dyke dolerite is made from magma superheated and pressurised before being injected up into the earth's crust. The fine grain of the dolerite (compared with, for example, granite) shows that the dyke cooled very quickly, as we would expect in such a narrow fissure. The photos above show how the surrounding sedimentary rocks were baked by the injection of hot magma.

A covering of snow picks out the landforms around Newtondale gorge. Here in its upper reaches the valley has a wide flat bottom and a definite meandering structure; both clues to its origin as a channel for glacial meltwater.

Newtondale

One of the most spectacular landscape features of the region, Newtondale cuts a deep and winding gorge through the southern part of the moors. It is a stunning example of a glacial outflow channel – a waterway carved by the run-off from ice age lakes and ice sheets. The keys to Newtondale's history are its narrowness and depth, which show that it is a recent landform, and the way that the valley winds with the stream, which shows that it was carved very rapidly.

Though rivers often take a winding course, their valleys are generally straightened out over time. Newtondale gorge itself winds in a series of open meanders. This shows that the gorge was formed rapidly by the sudden outflow of a large amount of water, cutting its way down through the plateau of the moors. The waters came from the drainage trapped in upper Eskdale. During and after the last ice age there were vast quantities of water trapped in ice sheets and glacial lakes – and all this water had to go somewhere.

◀ The meltwater overflowing from Lake Eskdale and the North Sea ice sheet cut a channel through the thick layer of moorland grit. The gorge begins at Fen Bog and winds its way south all the way to Pickering. This view shows the gorge running from Fen Bog towards the escarpment of the Tabular Hills, where it cuts a narrow route between Levisham Moor and Cropton Forest.

Avicula

Avicula echinata is a small bivalve, around 1 cm across. Fragments of its shells are commonly found in Cornbrash sediments.

◀ The steep route from Levisham down to Levisham Station reveals quarries of Lower Calcareous Grit and a view of the gorge winding ever south.

NYM Railway

The gorge was used by George Stephenson for part of the route of the Whitby to Pickering railway in 1836. Newtondale now carries the North Yorkshire Moors Railway, which runs steam and diesel trains between Pickering and Grosmont, and on to Whitby.

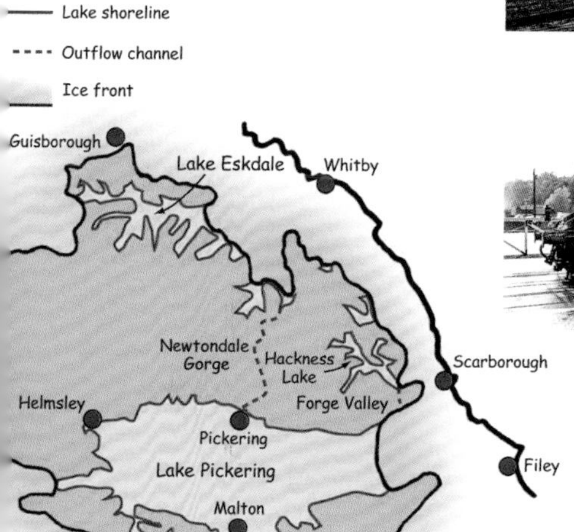

▲ Stations at Goathland and Levisham are a distance from their villages in order to meet the rail route at the bottom of the gorge. In contrast Grosmont village grew up around the railway as a vibrant industrial and transport centre in the nineteenth century.

Rosedale

The high moorland is in the form of an arch running from west to east. A series of magnificent dales runs down either side of this arch. To the north the dales run into Eskdale and to the south they run as far as the Tabular Hills, where they are dramatically altered by the change in geology. The southern moorland dales include Bransdale, Farndale and Rosedale. Here we look in detail at Rosedale – now a peaceful dale but once, due to its geology, a hive of industry.

▲ Low evening light at Rosedale picks out the course of the old railway line on the eastern side of the dale. The ironstone was mined from the Dogger formation. Fertile pastures below the railway are a sign of Lower Jurassic shales.

Ironstone mining

Like the other moorland dales, Rosedale has been cut through the hard crust of moorland grit into the soft Lower Jurassic rocks beneath. Those rocks include a thick layer of Dogger, and for several hundred years Rosedale had a thriving ironstone industry. The ironstone was mined on both sides of the dale and then transported on railways up either side of the valley. The routes of the railways are still highly visible.

The iron ore was tipped into kilns on both sides of the dale, for calcining. This process reduced the impurities in the ore and its weight. The calcined ore was tipped into wagons and taken across the moors.

▼ This photograph shows the double rail line on the east side of the dale, with ore tipped into calcining ovens and then tipped again for transport out of the dale.

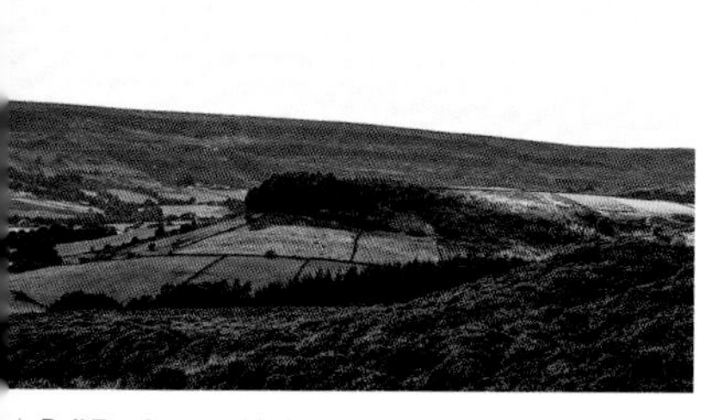

Bell Top forms a kind of headland in the middle of Rosedale. The boundary between the moorland grit and Whitby Mudstone is marked by the dramatic change from heather moorland to pasture.

Arches on the top of Chimney Bank show the rail route and the calcining ovens of the west side of the dale.

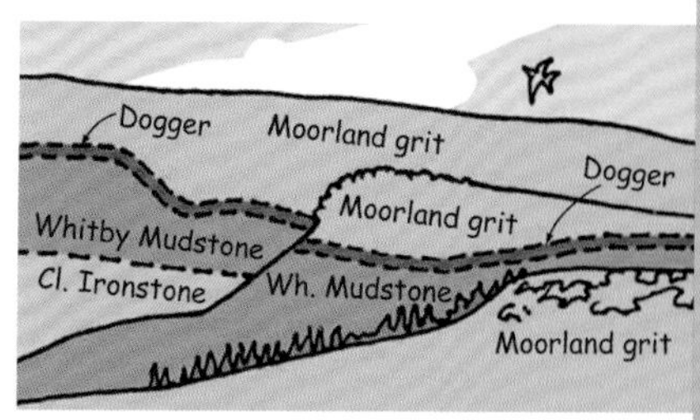

This section through Bell Top shows the geological boundaries reflected in the vegetation. While the ironstone in the Cleveland and Eston Hills was mined from Cleveland ironstone, in Rosedale a rich seam of Dogger was the source rock.

The route along the east side of Rosedale shows the massive engineering involved in building the railway, with embankments and cuttings still clearly visible.

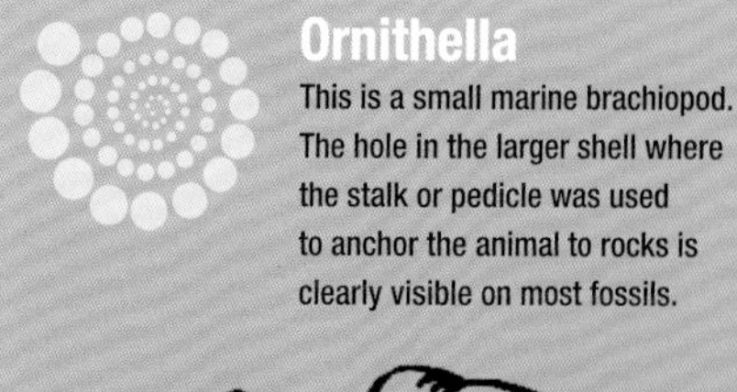

Ornithella

This is a small marine brachiopod. The hole in the larger shell where the stalk or pedicle was used to anchor the animal to rocks is clearly visible on most fossils.

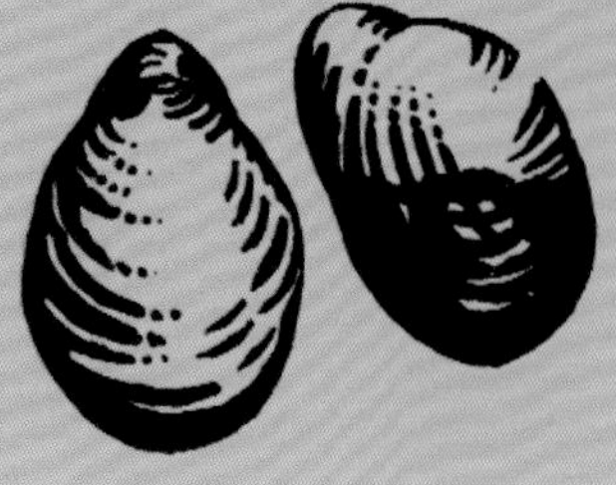

Hole of Horcum and Levisham Moor

The Hole of Horcum is the best-known landform on the North York Moors. Every year tens of thousands of walkers are drawn by the huge green bowl that seems to have been scooped out of the surrounding moorland and forest. With heather moorland above and pastures below, the Hole of Horcum gives us clues to the rocks that lie beneath, while the gulleys and streams that flow down its sides give an indication of how it was formed. Just to the west lies the Levisham estate, a unique landscape owned and managed by the National Park, which also owes its character to the underlying geology.

These views of the Hole of Horcum show the different types of vegetation on the slopes. On the left (the eastern side) the upper slopes are covered in trees. This area is underlain by Lower Calcareous Grit. Beneath this is a zone of pasture land which sits on top of Oxford Clay. On the other side, the tops and slopes have been cultivated differently. Here the grit on the top has been used for heather moorland, regularly burned off to bring new shoots – and to prevent the growth of tree cover. The lower slopes are a mix of heather and invasive bracken. On the west side of the beck is moorland in contrast to the pastures on the east side. The floor of the Hole of Horcum is underlain by Oxford Clay and Osgodby sandstone. This interbedding of sandstone and clay supports pasture and moorland fed by springwater from the slopes above.

◀ The north and west slopes are a spectacular sight from the main road and carpark. The bowl of the Hole of Horcum has been formed by a series of springs acting on the underlying rock.

Springs

Hidden among the heather, bracken and trees on the upper slopes are a series of deep gullies, or griffs, that carry water down to the beck below. Gulleys on the slopes are clearly visible. Although difficult to see from above, the griffs are obvious to anyone walking down the slopes of the Hole of Horcum and these are the key to its formation. As water soaks through the calcareous sandstone on the tops it reaches the next layer of rock, which is the Oxford Clay. The clay is impervious to water, and so the water finds its way out as a series of springs at the junction of the two layers. The gradual action of springwater on the soft clay has created this spectacular landform.

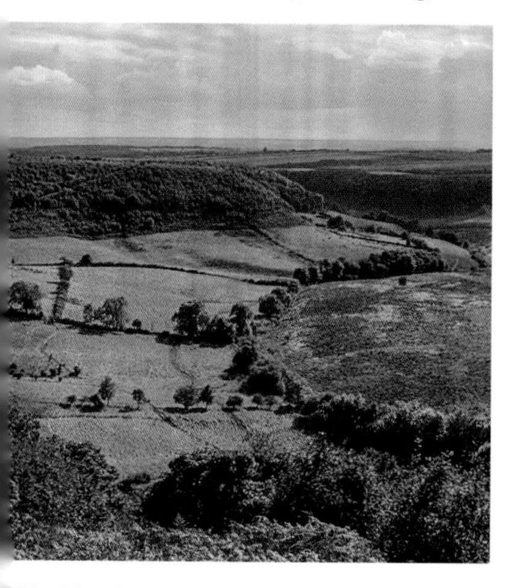

◀ The floor of the Hole of Horcum shows Levisham Beck gathering the springwaters and flowing south.

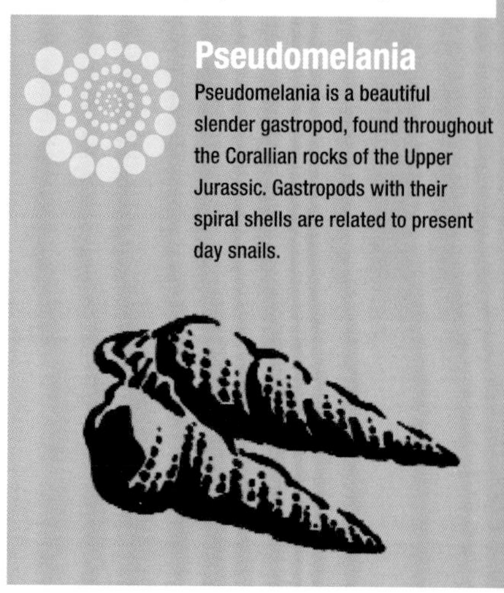

Pseudomelania

Pseudomelania is a beautiful slender gastropod, found throughout the Corallian rocks of the Upper Jurassic. Gastropods with their spiral shells are related to present day snails.

▼ Levisham Moor extends from the Hole of Horcum west to Newtondale. This area contains a unique combination of remains of human habitation, wildlife and vegetation and is legally protected. The moor is part of the geological structure of the Tabular Hills. The photo below shows the two levels of the moor. The upper shelf is Lower Calcareous Grit, while the lower is Osgodby Formation sandstone.

Row Brow above Scarborough is part of the long escarpment of the Tabular Hills that reaches all the way to Helmsley. This landform is seen across the region with hard rock at the top supporting forest and the lower clay slopes supporting pasture fields.

The Tabular Hills

The Tabular Hills run from west to east along the southern part of the National Park, all the way from Bilsdale to the coast at Scarborough. The underlying geology of these hills produces a beautifully varied landscape including dry valleys, wooded dales, open pasture land, forests and, on the northern margin, a spectacular escarpment.

◀ Open arable fields like these above Ebberston sit on lime-rich rocks. To th north Dalby Forest, li Cropton, Langdale ar Broxa forests, lies on infertile calcareous g

▼ The Tabular Hills were heavily quarrie for building stone bu also for limestone to be baked in kilns for fertiliser.

The Tabular Hills are made from rocks of Upper Jurassic age, which sit on top of the Middle Jurassic rocks that form the great plateau of the central moorland. These Upper Jurassic rocks are a mixture of limestones and lime-rich, or calcareous, sandstones. The high limestone content makes them quite different from the rocks of the central moorland. The lime-rich soil supports a range of wild plants that flourish on the verges and pathways of this landscape.

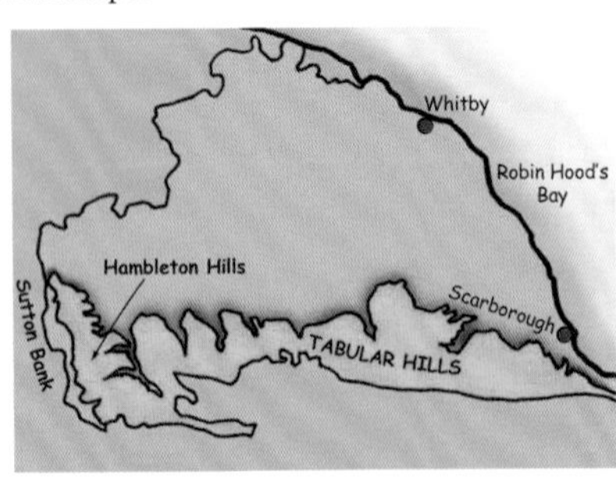

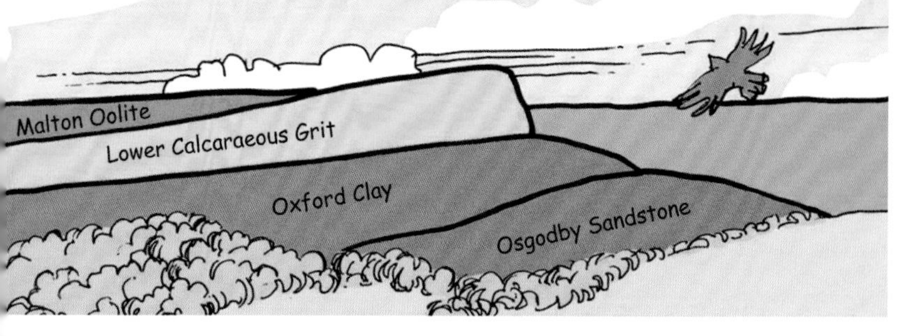

Dry valleys are a common feature of his area. Carved by glacial meltwater, the nderlying limestone is porous so water ow flows beneath the ground.

Water and limestone

The rock beds of the Tabular Hills dip gently towards the south. The lime-rich strata soak up rainwater which is carried through the strata towards the Vale of Pickering. This means that there is no running water on the upper reaches of the hills. The few villages on the tops – Fadmoor, Gillamoor, Newton on Rawcliffe, Broxa, Silpho – all built large ponds and in some cases towers to store water. In contrast, where the limestone meets the clay of the Vale of Pickering there are numerous springs to feed the villages along the A170.

The water tower at Silpho is an indicator of the porous limestone beneath the village.

The Tabular Hills are noted for their nestone flora, including cowslips and chids.

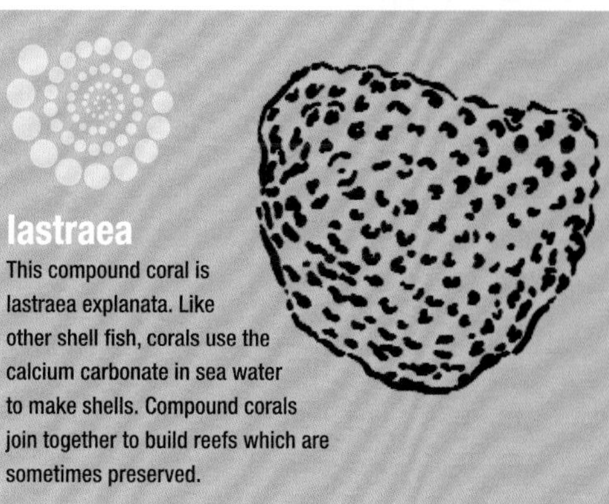

lastraea

This compound coral is lastraea explanata. Like other shell fish, corals use the calcium carbonate in sea water to make shells. Compound corals join together to build reefs which are sometimes preserved.

Southern dales

The dales that run south from the high moorland widen out into beautiful green landscapes. But when they reach the Tabular Hills their shape is altered by the change in geology. The beck waters have cut through these hills to create a series of wooded dales that are quite different to the dales further north. Elsewhere the porous limestone hills have few streams, while on the southern fringe of the hills a layer of impervious clay creates springs.

▲ The wooded slopes of the Tabular Hills dales, such as Douthwaite Dale, are a complete contrast to the moorland dales.

Douthwaite Dale is in effect the southern extension of Farndale, carrying the waters of the River Dove through the Tabular Hills. Here the dale is a narrower valley with wooded slopes. Further south the beck runs into Kirkdale as the dale opens out with the stream slowly meandering along the flat valley floor.

The Dove and the other becks have worn their way through the layers of rock in the Tabular Hills. A typical section through one of these dales (below) shows the layers of rock sloping down to the south. The land roughly follows this dip but has a more shallow slope.

▲ Hodge Beck at Kirkdale flows over a limestone platform. Sink holes take much of the water into an underground aquifer.

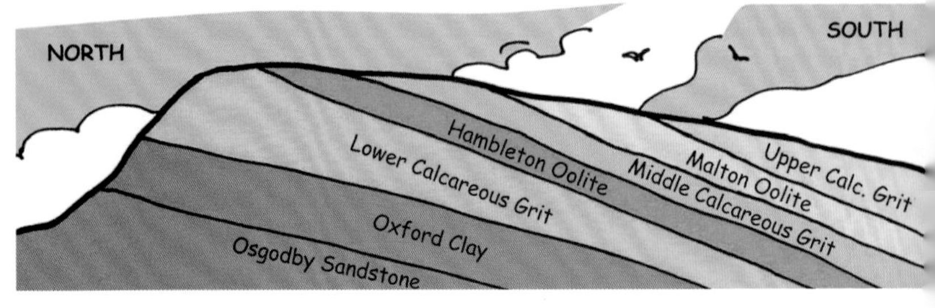

◀ Old limestone quarries are a common sight on the Tabular Hills.

▲ The Bridestones are near Staindale Lake in Dalby Forest; they are part of a National Trust property that includes the outlier at Blakey Topping. The stones are natural sculptures probably formed during the last ice age.

Kirkdale Cave

Near to the Kirkdale ford is a quarry which contains one of the most important sites in the history of geology. In 1821 workmen found hundreds of bones of ice age mammals in the floor of this cave (above). William Buckland analysed the material and suggested that this had been a den for hyenas who dragged the carcases of dead animals into their cave. The bones include rhinoceros, mammoth, bison, birds and hyenas. Buckland's study deduced the behaviour of the animals and their place in the surrounding environment, and ushered in the new science of palaeontology – literally 'the study of ancient life'.

Perisphinctes

Ammonites are common in the Upper Jurassic marine sediments. Perisphinctes ammonites thrived in these shallow coral seas.

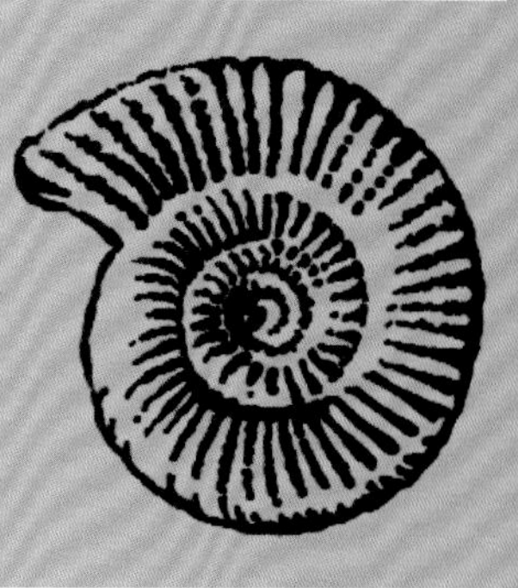

▼ Dales emerge from the Tabular Hills on to the Vale of Pickering in a series of wooded valleys. This landform at Wyedale near Snainton is replicated at intervals along the main road from Ayton all the way to Helmsley.

Forge Valley and Hackness

In the far southeast of the National Park lies one of the jewels in the crown of the North York Moors. Forge Valley is a National Nature Reserve, one of the most important mixed woodlands in the north of England, and an enchanting landscape with lots of geological exposures. The valley was cut through the Tabular Hills during and after the last ice age, by water running off from a lake covering present-day Hackness. The lake was held in place by the North Sea ice sheet. When this melted it left behind a bank of mud that diverted streams away from the sea. The Derwent now flows down Forge Valley and through the Vale of Pickering before its waters eventually join the Humber.

River Derwent

The Derwent should really flow eastwards to the sea but has been diverted down Forge Valley by glacial mud. The steep sides of the valley show that it is a recent creation, cut by glacial meltwater. Just as the river flows out of the valley near West Ayton much of its water disappears down through the dipping layers of rock into an underground aquifer. This aquifer is used to supply Scarborough – a town without a major river – with its water. Just as the river flows out of the valley near West Ayton.

◀ **Whetstone quarry is at the north end of Forge Valley high up on the eastern hill side. The Lower Calcareous Grit here is a deep yellow. As the name implies the stone here was quarried for grinding wheels, used for sharpening knives and tools.**

Glacial outflow

The waters that cut Forge Valley out of the Tabular Hills flowed out of an ice age lake that sat in the bowl of the hills around Hackness. Below 150 metres the land slopes very steeply down to a flat floor. The lake was dammed on the eastern side by the North Sea ice sheet, leaving the waters to flow south towards Forge Valley. The Sea Cut was built to relieve flooding.

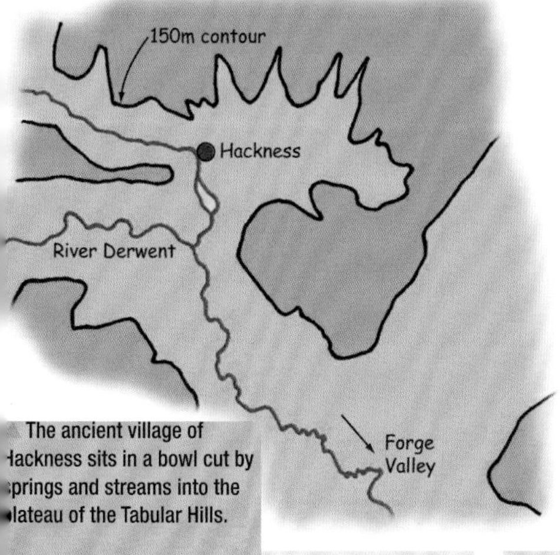

The ancient village of Hackness sits in a bowl cut by springs and streams into the plateau of the Tabular Hills.

Located at the southern end of Forge Valley, Whitestone Quarry is an exposure of Hambleton Oolite, one of the highest beds in the Upper Jurassic succession. Notice the difference in colour between the yellow sandstone (opposite) and the white oolite.

The Derwent has cut a flat-bottomed valley between Hackness and Forge Valley.

The steep road up from Hackness to Silpho offers views of the village and the Tabular Hills escarpment.

Nilsonnia

The Middle Jurassic rocks of Yorkshire are famous for their plant fossils. Conifers, tree ferns, gingkos ferns and cycads, such as this Nilsonnia grew across the region. Leaves and stems have been preserved in the sand and mudbanks.

The view of Robin Hood's Bay from Ravenscar shows the rings of Lower Jurassic strata that form the graceful curve of the bay. The Yorkshire coast is famous for its geology and the abundance and diversity of its fossils.

Cliffs and coves

The cliffs of the North Yorkshire coast are made of layers of Jurassic rock towering above the North Sea. From Hunt Cliff at Saltburn in the north to Filey Brigg in the south, an almost unbroken succession of cliffs is interrupted only by small coves and the occasional sandy beach. The cliffs are so high and steep because the same hard rocks that form the plateaux of the high moors and the Tabular Hills also make the tops of the high cliffs.

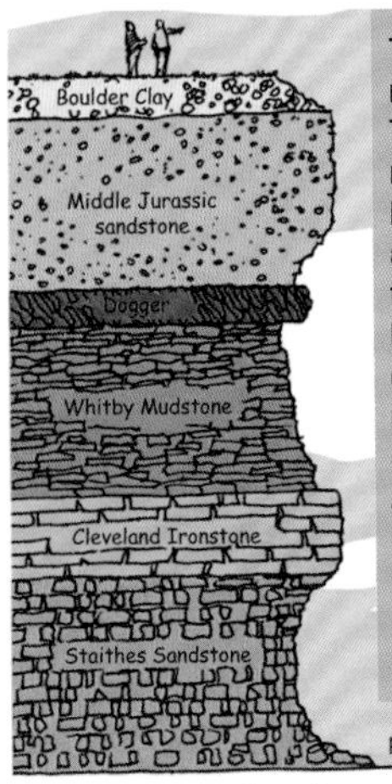

This section of cliff is from the north of the Yorkshire coast. The hard layer of sandstone is resistant to erosion, but the soft layers beneath are being eaten away by the sea. The result is that huge lumps of sandstone break off and tumble down. This combination of hard cap and soft underbelly is what makes the cliffs high and steep.
Above the sandstone is a layer of boulder clay. This is a relic of the last ice age, left behind when the ice sheets melted around 11,500 years ago.

Industrial coast

Beneath the Middle Jurassic clifftc of much of the coast there are huge exposures of Lower Jurassic sediments. These strata contain mineral riches including alum, jet and ironstone that were heavily exploited in the past. There are ol alum quarries and works at Boulb Kettleness, Sandsend, Saltwick and Ravenscar, jet holes and mine wherever there are exposures, and ironstone docks at Port Mulgrave. The rocky scars are crossed by ol tramways and the cliffs have been moulded by quarrying.

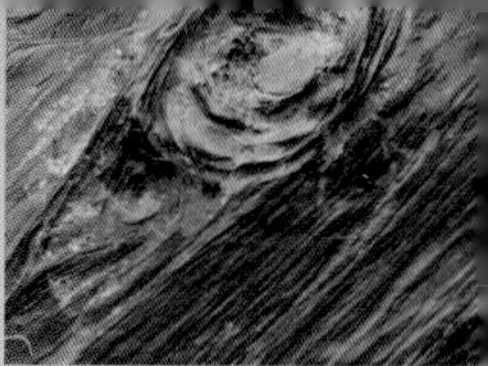

Jet is a fossilised wood that occurs in a particular bed within the Whitby Mudstone Formation. Jet is made out of the fossilised trunks of monkey-puzzle trees that grew along the coast, and were washed into the sea when they fell. The wood of the trees was fossilised and preserved in the shales forming at the bottom of the Jurassic sea. When cut and polished, jet attains a beautiful deep lustrous black colour; Whitby jet is prized all over the world.

▲ The northern cliffs show layers of the Cleveland ironstone formation laying almost horizontal above Staithes sandstone.

▲ The cliffs on the east side of Whitby show hard sandstones overlying the softer Whitby Mudstone with a layer of Dogger in between.

▼ The strata at Robin Hood's Bay are the remnants of a dome structure that has been worn away by erosion.

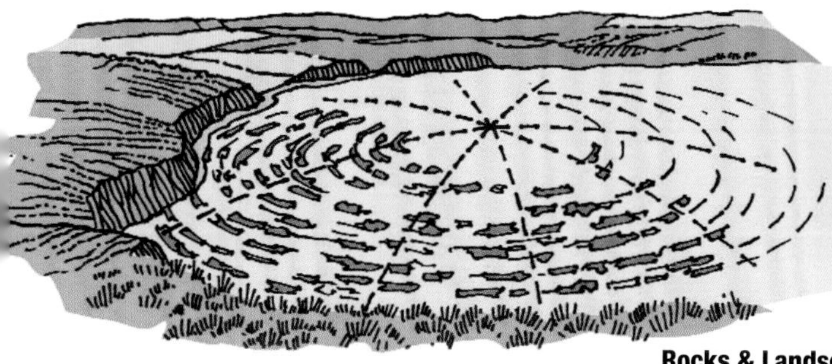

Coastal plain

Between the high cliffs of the coast and the heather moorland lies a captivating strip of undulating green pasture land. The pasture fields lie on top of glacial till deposits left behind by the North Sea ice sheet when it melted around 11,500 years ago. The ice sheet pushed in as far as the high ground of the North York Moors, suggesting that there was already some kind of coastal plateau. The glacial clay has produced soil that is fertile enough for rich pastureland and some arable farming.

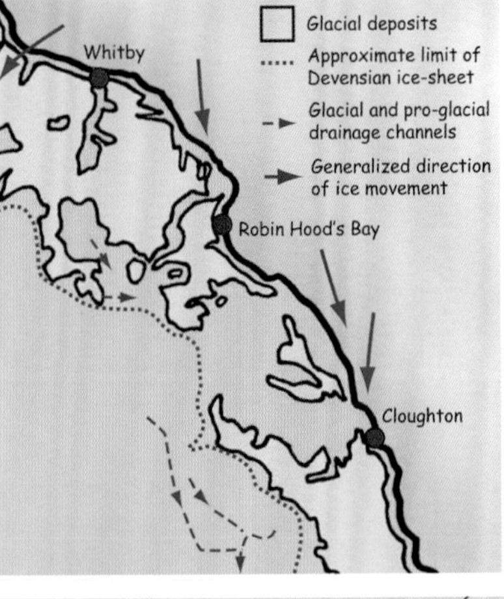

Ice age remnants

The landscape along the coastal strip is gentl[y] rolling pasture hills with fields, hedges, stream[s] and copses. The map on the left shows the extent of the glacial deposits along the coast and the direction of travel of the ice sheets. Note how the till does not extend over the hi[gh] land of the moors.

The till has had a dramatic effect on drainag[e]. Streams that wound their way to the coast were blocked and had to find new routes. The villages at Staithes, Runswick and Robi[n] Hood's Bay are steep because the streams that feed them are recent, and have cut deep narrow gulleys. You will see blocked channel[s] at Saltwick, Sandsend and Cayton Bay.

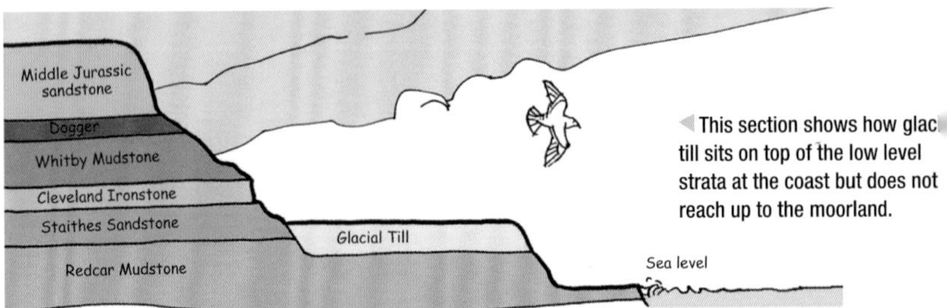

This section shows how glac[ial] till sits on top of the low level strata at the coast but does not reach up to the moorland.

The pebbles on Yorkshire's beaches owe heir huge variety to the ice sheets. A whole ange of rocks were brought to the region rom the north by ice. These large pebbles t Cloughton Wyke have been rounded off y the continuous action of the sea.

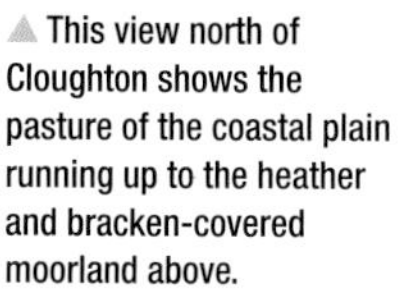

This view north of Cloughton shows the pasture of the coastal plain running up to the heather and bracken-covered moorland above.

Erratics like this large boulder at Crossgates were brought onto the coast and inland by ice sheets.

The Mere at Scarborough shows glacial till blocking drainage.

Mammoth tooth

Remains of the animals that inhabited the region during the ice ages show that the environment was tundra. Mammals like reindeer, wolverine, fox, bear and hare were here, along with mammoth and woolly rhinoceros. This mammoth tooth was found at the coast near Flamborough Head, just south of our area. Ice age remains have not had time to become fossilised, instead the bones are preserved in mud.

The Mere lies in a valley near Scarbrough that cuts off Oliver's Mount from the rest of the Tabular Hills. The valley proably carried a substantial water course before the last ice age. Now it is full of glacial till and the waters are trapped in the tranquillity of the Mere.

Useful information

We hope this book will encourage you to explore the landscape of the North York Moors National Park and the surrounding area. Whitby Museum has spectacular displays of local fossils:

Whitby Museum, Pannett Park, Whitby
www.whitbymusuem.org.uk

There are lots of places to get further information about the National Park.
The main National Park information centres are at:
Sutton Bank National Park Centre
01845 597426
The Moors Centre, Danby
01439 772737
The North York Moors National Park website is at www.northyorkmoors.org.uk

Further reading

High Tide Publishing's books are available from bookshops and other retailers across north-east Yorkshire:

The Dinosaur Coast: Yorkshire Rocks, Reptiles and Landscape (2015)
Beach Finds on the Yorkshire Coast (2016)
Fossils of the Yorkshire Coast (2018)

Check out these related titles:
Roger Osborne (2006) *Discover the North York Moors*, North York Moors National Park
Roger Osborne (1998) *The Floating Egg*, Jonathan Cape

The British Geological Survey publishes geological maps of the area, available for purchase via their website. You can also look at the geology of the area online through the BGS Geology of Britain Viewer. Visit www.bgs.ac.uk and follow the links.